For Grade **3**

This Book Belongs To:

Read on Target

Using Reading Maps to Improve Reading Comprehension and to Increase Critical-Thinking Skills

Written By:
Sheila Anne Dean, M.S., CCC-SLP
Jeri Lynn Fox, M.S.
Pamela B. Meggyesy, M.A.
Pamela Marie Thompson, M.S.

Published By:
Show What You Know® Publishing
A Division of Englefield & Associates, Inc.
P.O. Box 341348
Columbus, OH 43234-1348
1-877-PASSING (727-7464)

www.showwhatyouknowpublishing.com

Copyright © 2005 by Englefield & Associates, Inc.

All rights reserved. No part of this book, including interior design, cover design, and icons, may be reproduced or transmitted in any form, by any means (electronic, photocopying, recording, or otherwise), without the prior written permission of the publisher.

Printed in the United States of America
20 19 18 17 16 15 14 13 12 11 10 9 8 7 6 5 4

ISBN: 1-59230-129-0

Limit of Liability/Disclaimer of Warranty: The authors and publishers have used their best efforts in preparing this book. Englefield & Associates, Inc., and the authors make no representation or warranties with respect to the contents of this book and specifically disclaim any implied warranties and shall in no event be liable for any loss of any kind including but not limited to special, incidental, consequential, or other damages.

About the Authors

Sheila Anne Dean, M.S., CCC-SLP, received a Bachelor's Degree from Ohio University and a Master's Degree from Miami University. She has worked for more than ten years in the public schools as a speech pathologist and a speech pathologist supervisor. Her role as a speech pathologist includes collaboration with teachers, parents, and other support staff while working with students individually, in small groups, and in the classroom setting. Previous presentations have included incorporating reading skills into therapy for school success, aligning instruction to meet classroom expectations, designing effective IEPs, and enhancing communication skills for at-risk and disabled children.

Jeri Lynn Fox, M.S., holds a Bachelor of Arts Degree from Bluffton College and a Master of Science Degree from the University of Dayton. A Martha Holden Jennings scholar, she has worked in public education for more than twenty-five years as a classroom teacher and school counselor. In her current role as school counselor, her responsibilities include improving student achievement, assessment coordination, and developing supportive parent/teacher partnerships and programming.

Pamela B. Meggyesy, M.A., earned a Bachelor of Science Degree in Education and a Master of Arts Degree in literature from Ohio University. Additionally, she has studied at Oxford University in Oxford, England. A thirty-year veteran classroom teacher in public schools, she has been involved with curriculum alignment and literature selection at the district and county levels. Additionally, she has chaired a district-wide writing initiative for her school system. She also has been an instructor at Wright State University.

Pamela Marie Thompson, M.S., has a Master's Degree in counseling from Wright State University and a Master's Degree in school psychology from the University of Dayton. She has taught at the college level and worked as a counselor and program director for a mental health clinic where she supervised case managers and a partial hospitalization program. She is currently employed as a school psychologist. Her background includes experience in testing, assessment, and academic intervention. She has worked in this field for eighteen years. An often sought-after speaker, she has presented numerous workshops dealing with a variety of educational topics including program evaluation, reading comprehension skill development, behavioral interventions, problem-solving skills, and instructional planning for the inclusive classroom. She has been a presenter for the Ohio School Psychologists Association state conference, the Association of School Administrators in Washington State, and the International Reading Association.

The authors have worked together for more than a decade in the Northridge School District in Dayton, Ohio. Their varied educational backgrounds and experiences bring a multifaceted approach to their collaborative educational projects.

© 2005 Englefield & Associates, Inc.

Acknowledgements

Show What You Know® Publishing acknowledges the following for their efforts in making this assessment material available for Ohio students, parents, and teachers.

Cindi Englefield, President/Publisher
Eloise Boehm-Sasala, Vice President/Managing Editor
Lainie Burke Rosenthal, Project Editor/Graphic Designer
Erin McDonald, Project Editor
Rob Ciccotelli, Project Editor
Christine Filippetti, Project Editor
Jill Borish, Project Editor
Jennifer Harney, Illustrator/Cover Designer

Dedication

To our families:

David, Sarah, Marie, Elaine, and Erin, thanks for your love.
Because of you, I truly "enjoy every day." –SAD

Richard, Amy, and Anna, thank you for the joy you bring to my life. –JLF

Joe, Mark, and Lauren; I treasure each of you. –PBM

Nathan, Olivia, and Jason, thanks for your love and encouragement. –PMT

and

To our first teachers:
our parents, by birth and by marriage,
with love and appreciation:

George and Carolyn Harrington and Carl and Marianna Dean –SAD

Charles and Edith Harlow and Richard and Janice Fox –JLF

George and Connie Besuden –PBM

Joe and Emily and Harold and Marie –PMT

© 2005 Englefield & Associates, Inc.

Table of Contents

Introduction	vii
Activity 1: Analyze Aspects of the Text by Examining Characters	1
Activity 2: Analyze Aspects of the Text by Examining Setting	6
Activity 3: Analyze Aspects of the Text by Examining Plot	12
Activity 4: Analyze Aspects of the Text by Examining Problem/Solution	17
Activity 5: Analyze Aspects of the Text by Examining Point of View	22
Activity 6: Analyze Aspects of the Text by Examining Theme	27
Activity 7: Infer from the Text	31
Activity 8: Predict from the Text	39
Activity 9: Compare and Contrast	47
✓ Activity 10: Analyze the Text by Examining the Use of Fact and Opinion	55
Activity 11: Explain How and Why the Author Uses Contents of a Text to Support His/Her Purpose for Writing	67
Activity 12: Identify Main Idea/Supporting Details	75
Activity 13: Respond to the Text	83
Activity 14: Evaluate and Critique the Text	91
Activity 15: Summarize the Text	99
Activity 16: Identify Cause and Effect	112
Self-Scoring Chart	127

Introduction

What is *Read on Target*?

Read on Target is a book that has 16 reading maps to help you answer tough questions related to something that you have read. *Read on Target* gives students, like you, the tools you need to answer critical-thinking skill questions. Some of these skills include the ability to analyze the story elements, to infer, to predict, and to compare and contrast. You will also need to know how to answer questions like analyze fact and opinion, explain how the author uses contents of the text to support his/her purpose of writing, critiquing, evaluating, summarizing, and determining cause and effect. These are the important thinking skills that are found in this book. Many of these skills teach you how to break down information and to show the relationships in the text. The reading maps in *Read on Target* will guide you, step by step, through the process of answering these types of questions. Each and every reading map is designed to help your ability to reason and to understand what you read.

Why do you need *Read on Target*?

Sometimes when teachers or parents want to check your understanding of what you have read, they will want to know if you can reason and think critically about information. If you can answer these types of questions, you can show what you've learned. Sometimes it's hard to come up with the right answer when you don't know what they're asking or how to answer the question. *Read on Target* will help you understand how to answer these questions. This book will also help you answer tough questions found on many of your tests. By using this book, you'll be more prepared after reading to answer those questions you know your teachers or parents will ask.

Reading and understanding what you have read are two difficult tasks that you are expected to do for every subject. So, when you know how to read and how to understand the text, you can participate in class tasks better and answer those tough questions. You will continue to use these thinking skills as you enter the world of work.

How do you use *Read on Target*?

Read on Target will guide you through the process of answering critical-thinking questions correctly. *Read on Target* will tell you what to look for when you are reading. You will write your answers on the reading map and use your text as you need it. It also helps you practice in your class and lets you see how someone else came up with a good answer. The next time you are asked to answer a critical-thinking question, you will be more prepared and better-equipped to provide a complete, well-thought out answer.

Activity 1

Analyze Aspects of the Text by Examining Characters

I read to figure out what the characters are like. I get to know them.

Step 1 Read the story "Sugar and Spice."

Sugar and Spice

Everyone I know has a younger brother or sister. No one has a little sister quite like mine! Her name is Erin. She is almost four years old, but she thinks she is ten. Even though she drives me crazy, I love her very much.

One time when we were playing, I heard her say, "Sugar and spice and everything nice. That's what little girls are made of!"

She is made of sugar all right. Almost everything about her is sweet. Her eyes are deep brown and sparkle in the light. Her skin is soft and smooth. Her dark brown hair shimmers. Her cheeks are so round that people sometimes say, "Look at her cute cheeks."

Erin is sweet like sugar, but she has a "spicy" side, too. She loves to laugh and make other people laugh. She likes jokes. She likes to play tricks on people. One time, she put a fake spider on Mom's books. She laughed and laughed!

She thinks it is fun being with her family and friends. She likes to do whatever everyone else is doing. If they are skating, she wants to skate. If they are writing, she wants to write. If they are playing outside, she wants to play outside. She thinks she is a big girl, too.

© 2005 Englefield & Associates, Inc.   COPYING IS PROHIBITED   1

Activity 1   Read on Target for Grade 3

She feels sad when people say mean things. She likes to be kind. Sometimes, she says the nicest things when we are together. One time she said, "I like being your sister."

I said, "I like being your sister, too!"

Then she said, "I really mean it!"

I said, "I really mean it, too!"

I think she's right about being made of sugar and spice.

Step 2 Student Tips

To analyze a character, you need to remember:

- A character can be a person, an animal, or an object.
- What the character is like, because this affects the story.
- The story could change if you change one part of a character.

Step 3 Complete the reading maps. Use the reading maps to help you think about the character.

Activity 1 *Read on Target* for Grade 3

Analyze Aspects of the Text by Examining the Characters

I read to figure out what the characters are like. I get to know them.

Map 1.1

Character's Name: _____

Describe the character.	Write a sentence from the story that tells about the character.	What does this tell you about the character?
What does the character look like?		
How does the character act?		
How does the character feel or think?		
What does the character say?		

© 2005 Englefield & Associates, Inc.   COPYING IS PROHIBITED   3

Activity 1 — Read on Target for Grade 3

Map 1.2

Analyze Aspects of the Text by Examining the Characters

I read to figure out what the characters are like. I get to know them.

Change the Character.

Change the character to **a mean girl**.
Tell how the story would be different.

_____
_____
_____
_____
_____

Change the Character.

Change the character to **a shy girl**.
Tell how the story would be different.

_____
_____
_____
_____
_____

Activity 1   Read on Target for Grade 3

Step 4 **Read the following questions and write your answers.**

1. What kind of person is Erin? Give an example from the story of something she does to support your answer.

   _____

   _____

   _____

2. Erin likes to make people laugh. What does this tell you about Erin?

   _____

   _____

   _____

3. In the story, it says that Erin is made of sugar. Tell one way that you know she is sweet.

   _____

   _____

   _____

4. How would the story be different if Erin were a mean girl?

   _____

   _____

   _____

© 2005 Englefield & Associates, Inc.   COPYING IS PROHIBITED

Activity 2

Analyze Aspects of the Text by Examining Setting

I figure out how important the setting is and how the setting affects the characters and events that take place.

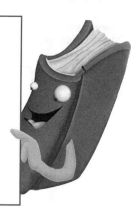

Step 1 Read the story "The Fire Escape."

The Fire Escape

Teddy heard the sirens screaming through the city streets. He stuck his head out his bedroom window. The window opened onto the fire escape. It was still warm outside even though the sun had gone down, and it was almost dark. He could feel the warm, moist air through the open window. The city lights were blinking red, neon pink, and orange. They seemed to announce the stores and businesses all around his apartment building.

School had been out for nearly a month. His mother didn't want him to be out after dark. "I want to know where you are," she said. Teddy was getting bored, wishing he could go out. Aunt Aggie had stopped by about an hour ago to make sure he was OK. She wasn't really his aunt, just the lady next door. She always looked in on him when his mom was at work.

Suddenly, he caught a whiff of something. It bothered his nostrils and his eyes. It was a burning smell. The sirens were getting louder. He saw people on the street below. They were yelling. They were looking up and yelling at him. They were saying, "Get out! It's your building!"

He could hear Aunt Aggie's television on next door. She couldn't hear very well, so her television was really loud. He ran to the door to let Aunt Aggie know there was a fire. The smoke smell was stronger by the door. Teddy ran back to the window.

The people on the street kept yelling. He knew he had to let Aunt Aggie know. He climbed out onto the fire escape. He looked down through the steel grating at the ground five stories below. He went to Aunt Aggie's window next door. He tapped really hard, but she still didn't hear him. He took off his shoe and hit the window with it. The glass shattered. Aunt Aggie screamed.

"Aunt Aggie, there's a fire! Come with me down the fire escape!" Aunt Aggie slowly pushed herself out of the chair. She followed Teddy out the window. It was hard for her. But with Teddy's help, she made it.

Together, they climbed down four stories of the fire escape. But the fire escape didn't go all the way to the ground. How would he and Aunt Aggie get to the ground?

Just then, the fire trucks pulled up. He could see a fireman running to call to them. "Just step on the ladder. It is spring-loaded and will let you get to the ground!"

"Thanks!" Teddy called to the fireman. Teddy and Aunt Aggie made it safely to the ground. Teddy guessed he wasn't so bored after all.

Activity 2   Read on Target for Grade 3

Step 2: Student Tips

To analyze the setting, you need to remember:

- What the setting looks like. Tell where the story takes place, tell when the story takes place, and tell what you hear, feel, smell, and see.

- The setting affects the story. If the setting is a sunny day, you might feel warm and happy. If the setting is a dark night, you might feel scared.

- The story could change if you change one part of the setting.

Step 3: Complete the reading maps. Use the reading maps to help you think about the setting.

Activity 2 *Read on Target* for Grade 3

Map 2.1

Analyze Aspects of the Text by Examining the Setting

I figure out how important the setting is and how the setting influences the characters and events that take place.

Describe the setting (where and when the story takes place).

	Write words or sentences from the text that tell about the setting.
Tell where the story takes place	
Tell when the story takes place	
Tell what you hear in the setting	
Tell what you feel in the setting	
Tell what you smell in the setting	
Tell what you see in the setting	

© 2005 Englefield & Associates, Inc.   COPYING IS PROHIBITED

Activity 2   Read on Target for Grade 3

Map 2.2 — Analyze Aspects of the Text by Examining the Setting

I figure out how important the setting is and how the setting influences the characters and events that take place.

Think about the setting.

Tell how the setting **(it is dark outside)** affects the characters.

_____
_____

Tell how the setting **(a fire escape)** affects the events of the story.

_____
_____

Now, change the setting.

Change what the setting looks like to **a one-story house**. Tell how the story would be different.

_____
_____
_____

Change where the setting is (where the story takes place) to **the country**. Tell how the story would be different.

_____
_____
_____

Activity 2 	 Read on Target for Grade 3

Step 4
Read the following questions and write your answers.

1. Where does the story take place?

   _____

   _____

   _____

2. Describe what you might see if you were with Teddy.

   _____

   _____

   _____

3. When does the story take place?

   _____

   _____

   _____

4. How would the story be different if the setting were a one-story house in the country?

   _____

   _____

   _____

© 2005 Englefield & Associates, Inc. COPYING IS PROHIBITED

Activity 3

Analyze Aspects of the Text by Examining Plot
I read to figure out the chain of events; what will happen next.

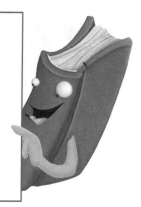

Step 1   Read the story "Forever Friends."

Forever Friends

On the first day of third grade, Anna became friends with Rosy. Anna and Rosy sat next to each other on the school bus. They sat together in class. They even sat together at lunch. The two friends were friendly with all of the third graders. But they especially liked to spend time together.

Anna and Rosy were riding their bicycles one Saturday morning in September. Anna's father called for the two friends to come to the edge of the sidewalk. He told the girls that he had just poured some cement to make a cement path near the flower garden. The cement was still wet. Anna's father had a fun idea for the two friends. Anna and Rosy pressed their hands into the cement. The sticky stuff was warm on their fingers. Anna found a stick, so the girls wrote their names next to their hands. Then, Rosy wrote "forever friends" next to their names. Finally, Anna drew a big heart around both of their names. In a few days, the cement was hard enough to ride their bikes on the new cement.

That spring, Rosy's family moved to a new house in a new city. Rosy said her family needed to live closer to her grandparents. Anna was very sad to say goodbye to her good friend. Anna had many other friends, but she really missed Rosy.

Then, a miracle happened. On the first day of fifth grade, Anna's mother told her there was a new girl moving into the neighborhood. Anna couldn't believe her eyes when she saw Rosy standing at the bus stop. Rosy's family had moved back to the neighborhood. Anna and Rosy were happy to be forever friends again.

Activity 3   *Read on Target* for Grade 3

Step 2 — Student Tips

To analyze the plot, you need to remember:

- The plot is the chain of events in the story. The plot has a beginning, a middle, and an ending. The plot has a problem and a solution.

- The plot affects the characters and events.

- When the plot changes, the story changes. What happens to the plot if you change the order of an event? What happens to the plot if you take out an event? What happens to the plot if you change the character's actions?

Step 3

Complete the reading maps. Use the reading maps to help you think about the plot.

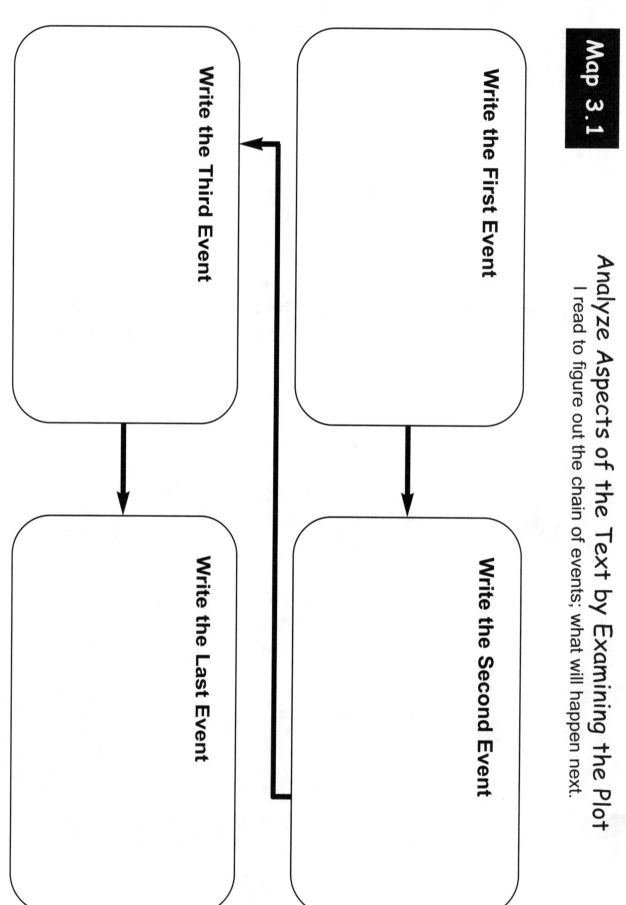

Activity 3 — Read on Target for Grade 3

Map 3.2

Analyze Aspects of the Text by Examining the Plot
I read to figure out the chain of events; what will happen next.

Change the Plot.

Change the event **(Anna became friends with Rosy on the first day of third grade)** to Anna and Rosy become friends in the spring of third grade. How is the story different when **this event happens later**?

_____
_____
_____
_____

Take the event **(girls making handprints and writing "forever friends" in the cement)** out of the story. How is the story different when **this event is left out of the story**?

_____
_____
_____
_____

What would happen if the **characters' actions** changed to **the girls not liking each other**?

_____
_____
_____
_____

Activity 3 Read on Target for Grade 3

Step 4
Read the following questions and write your answers.

1. What event happens at the beginning of the story?

   _____
   _____
   _____

2. What happens when the girls are riding bikes in September?

   _____
   _____
   _____

3. How does the story end?

   _____
   _____
   _____

4. How would the story be different if the girls didn't like each other very much in third grade?

   _____
   _____
   _____

Activity 4

Analyze Aspects of the Text by Examining the Problem/Solution

I read to figure out the problem and how it is solved.

Step 1 Read the story "Fox and Bear."

Fox and Bear

Fox and Bear are friends. Fox is quick and small. Bear is big and strong. They like to play in the big woods. Fox can run fast. Fox always wins the races with Bear. He can run under fallen trees and hide behind big rocks. Bear is too big to hide behind most rocks, so Fox always finds Bear when they play hide and seek. Bear is sad because Fox wins all of the games.

Fox has an idea. They will have a different kind of race. They will have a tree-climbing race. Fox can jump at the tree. His legs help him spring toward the tree trunk, but he cannot climb the tree. Bear is a very good tree climber. He has climbed trees since he was a cub. Bear wins the tree-climbing race. Fox and Bear are happy that each of them can win when they play different games.

Activity 4   Read on Target for Grade 3

Step 2 — Student Tips

To analyze the problem/solution, you need to remember:

- A problem from the story can be something the character wants to change or something the character wants to do.

- A solution from the story can be an action taken to solve the problem, or it can be a decision.

- The problem and solution help you understand the plot.

- If you change the problem, think about how the events from the story or the solution could change.

Step 3 — Complete the reading maps. Use the reading maps to help you think about the problem/solution.

Activity 4   *Read on Target* for Grade 3

Map 4.1 — Analyze Aspects of the Text by Examining the Problem/Solution

I read to figure out the problem and how it is solved.

Read the definition of a problem and a solution.

The **problem** can be:	The **solution** can be:
• A situation that the character wants to change.	• An action that helps the character understand how the problem is solved.
• Something the character wants to do or to find out.	• A decision that helps the character understand how the problem is solved.

What is the problem? _____

_____

_____

What events help solve the problem?

    Event 1. _____

    Event 2. _____

    Event 3. _____

What is the solution? _____

_____

© 2005 Englefield & Associates, Inc.   COPYING IS PROHIBITED

Activity 4 — Read on Target for Grade 3

Map 4.2 — Analyze Aspects of the Text by Examining the Problem/Solution

I read to figure out the problem and how it is solved.

Change the problem by making up a different problem.

The problem has changed to **Fox bragging about winning every game.**

How would the events be different?

Event 1. _____
_____

Event 2. _____
_____

Event 3. _____
_____

How would the solution be different?

_____
_____
_____
_____
_____
_____

Activity 4 *Read on Target* **for Grade 3**

Step 4

Read the following questions and write your answers.

1. What is the problem for Fox and Bear?

   _____
   _____
   _____

2. What events solve the problem for Fox and Bear?

   _____
   _____
   _____

3. How was the problem solved?

   _____
   _____
   _____

4. How would the solution be different if the problem changed to Fox bragging about always winning every game?

   _____
   _____
   _____

Activity 5

Analyze Aspects of the Text by Examining the Point of View

I figure out the author's choice of speaker. I also think about why the author chose to write from this point of view and how the story would be different if the story were told from another point of view.

Step 1 Read the story "A Good Citizen."

A Good Citizen

Grandma Cindy is a volunteer at my school. She knows every student by name. She makes our school library an exciting place. Every week, she reads stories to each class at my school. She teaches us about how to care for books. Sometimes, she tells us about the author or illustrator of a book. She also chooses interesting books for the display cases in our school library. Some of the books in the display are new, but many of them are older stories that she doesn't want us to miss. Every time our class reads one hundred books, Grandma Cindy brings cookies and punch to celebrate how much we are learning.

Grandma Cindy isn't really my grandmother, but everyone in my school calls her by that name. Even Mr. Kain, my principal, says, "What books have you selected for the library display today, Grandma Cindy?"

Grandma Cindy volunteers every day at our school. Before she came to our school, Grandma Cindy worked in a factory making cars. When she retired, she headed straight to our school and said, "I am here to help." Mr. Kain explained that there was no money to hire helpers for our school. But Grandma Cindy said she didn't need to be paid.

Grandma Cindy says being a volunteer at our school is one way she can be a good citizen. She wants to help support our school and make our community better. She says she loves being a volunteer. The kids at my school all want to be volunteers someday. Then, we can be good citizens like Grandma Cindy.

Activity 5   Read on Target for Grade 3

Step 2: Student Tips

- Make sure you know the definitions and key words of each point of view. You will find the definitions in the Reading Map.

- Think about the reason the author wrote from this point of view.

- Did the author write to let you know what was in the mind of several selected characters only, or did the author let you know what every character was thinking? Perhaps the author told the story to allow you to step into the shoes of the main character.

- Consider how changing the point of view will affect how you feel or think about what you have read. This will give you a clue as to why the author wrote from that point of view.

Step 3: Complete the reading maps. Use the reading maps to help you think about the point of view.

Activity 5　　　　　　　　　　　　　　　　　Read on Target for Grade 3

Map 5.1

Analyze Aspects of the Text by Examining the Point of View

I figure out the author's choice of speaker. I also think about why the author chose to write from this point of view and how the story would be different if the story were told from another point of view.

Point of View	Definition	Key Word Pronouns: They tell the author's choice of speaker.
First-Person	I am in the story. I tell the story.	I, me, my, we, us, our
Third-Person	Someone outside of the story tells the story from what he/she knows.	he, she, they, them
Omniscient (All-Knowing)	Someone outside of the story tells the story but knows what everyone sees, feels, and thinks.	he, she, they, them

	Write one or more sentences from the story that helped you figure out the author's point of view.
What is the Point of View? (circle who tells the story) 1. First-Person 2. Third-Person 3. Omniscient (All-Knowing)	

24　　　　COPYING IS PROHIBITED　　　　© 2005 Englefield & Associates, Inc.

Activity 5 | *Read on Target* for Grade 3

Map 5.2

Analyze Aspects of the Text by Examining the Point of View

I figure out the author's choice of speaker. I also think about why the author chose to write from this point of view and how the story would be different if the story were told from another point of view.

Why did the author write from this point of view?

Change the point of view to omniscient. How is the story different?

© 2005 Englefield & Associates, Inc. COPYING IS PROHIBITED 25

Activity 5   Read on Target for Grade 3

Step 4
Read the following questions and write your answers.

1. What is the point of view of the story? How did you figure out the point of view of the story?

   _____

   _____

   _____

2. How would the story be different if it was told in omniscient point of view?

   _____

   _____

   _____

3. Which point of view tells you about the characters from someone outside of the story?

   _____

   _____

   _____

4. What point of view would an author most likely use if the story was written to tell you how every character feels?

   _____

   _____

   _____

Activity 6

Analyze Aspects of the Text by Examining the Theme

I figure out the overall message the author is telling me.

Step 1

Read the story "Good Friends."

Good Friends

Sophie and Juan were good friends. One day, Sophie fell and scraped her knee during recess. Juan took her to the nurse's office. They had studied germs in science class and had learned to clean cuts. He thought that was the smart thing to do.

Sophie said, "Thanks Juan. I'm glad you helped me get my knee cleaned."

Juan said, "That's all right, Sophie. I was glad to help."

Another time, Juan did not know how to do his math homework. He tried and tried, but he just did not understand. His mother said, "Why don't you call Sophie to see if she could explain it to you? Maybe she can help."

Juan said, "That's a good idea, Mom." Juan called Sophie. She tried to explain it just like their teacher, Mrs. White, did. Sophie was able to help him. After a while, Juan understood the homework. Together they talked about the math steps. Then he was able to do the homework all by himself. Sophie really helped Juan.

During gym class, they had a relay race, and Sophie and Juan's legs were tied together. They had to walk across the floor to the other side. They talked about their plan. They held each other up when they wobbled. Each of them kept the other person from falling. They laughed. They shared. They understood what it takes to be a good friend.

Activity 6 Read on Target for Grade 3

Step 2: Student Tips

To analyze theme, you need to remember:

- There are important ideas from the story. Look for words that tell about the story. Look for repeated words.

- The character might learn lessons. Think about how the character feels and thinks. Think about what happens to the character.

- There is an overall message of the story.

Step 3: Complete the reading map. Use the reading map to help you think about the theme.

Activity 6   *Read on Target* for Grade 3

Map 6 — Analyze Aspects of the Text by Examining the Theme

I figure out the overall message that the author is telling me.

Name some of the ideas that you learned from the story.

_____
_____
_____
_____
_____

What lessons did the character or characters learn?

_____
_____
_____

Write a sentence telling what you think the theme of the story is.

_____
_____
_____

Activity 6   **Read on Target** for Grade 3

Step 4

Read the following questions and write your answers.

1. Write two ideas you learned from this story.

   _____

   _____

   _____

2. What does Juan think about helping Sophie? Use examples from the story to support your answer.

   _____

   _____

   _____

3. How does Sophie help Juan? Explain your answer.

   _____

   _____

   _____

4. What is the theme or important idea you get from this story?

   _____

   _____

   _____

Activity 7a

Infer from the Text

I read clues and use what I know to figure out what is happening in the story.

Step 1

Read the story "Hannah's Party."

Hannah's Party

Hannah wanted to get ready for her party. First, she wanted to decorate her pumpkin. She got a bucket to put the seeds in as she dug them out of the pumpkin. Hannah got a cloth and wiped off the pumpkin. She used corn for its ears and drew the eyes and nose in a round shape to make a face. Next, Hannah put a ribbon around her pumpkin. Hannah put her pumpkin outside with a sign that said, "Welcome."

Hannah thought that her friends would like to eat lots of treats. She made something very special. Hannah mixed flour, sugar, eggs, and water together to make dough. Then she rolled the dough flat and cut circles in the dough. Hannah baked the little dough circles. After they were baked, she put orange icing on them. Hannah thought the treats looked great.

After she made her treats, Hannah placed the orange plates and cups on the tables. She put the candy in the little dishes and put on her costume. Hannah put apples and a bucket on the table. She planned for everyone to bob for apples and play fun games.

Children started to knock at the door. Hannah was ready for her party to begin!

Activity 7a   *Read on Target* for Grade 3

Step 2 — Student Tips

To infer from the story, you need to remember:

- There are clues in the story. Clues are hints the author gives you about the story. Draw a line under each clue, or use your finger to point to the clue.

- What you know will help you figure out the story. Think about what is going on in the story. Have you done it before? Do you know about it?

- Clues, experiences, and knowledge are put together.

- Using clues from the story and what you know will help you figure out what is happening in the story. (This is called an **inference**.)

Step 3 — Complete the reading map. Use the reading map to help you think about inferring.

Activity 7a *Read on Target* for Grade 3

Map 7a

Infer from the Text

I read clues and use my knowledge to figure out what is happening in the story.

My Clues	+	My Experiences	=	My Inference
Find the clues in each sentence. Write your clues in the boxes below.		Think about your experiences and knowledge of a similar thing. Write your information in the boxes below.		Put the clue and experience/knowledge together to make a guess about what is happening in the story.

Clues to tell you the holiday.

→ Your experiences and knowledge of a similar thing.

⇒ What is the holiday?

Clues to tell you the treat.

→ Your experiences and knowledge of a similar thing.

⇒ What is the treat?

© 2005 Englefield & Associates, Inc. COPYING IS PROHIBITED 33

Activity 7a   Read on Target for Grade 3

Step 4
Read the following questions and write your answers.

1. What holiday is it?

   _____

   _____

   _____

2. What clues from the story helped you figure out your answer?

   _____

   _____

   _____

3. What treat did Hannah make?

   _____

   _____

   _____

4. What clues from the story helped you figure out your answer?

   _____

   _____

   _____

Activity 7b

Infer from the Text
I read clues and use what I know to figure out what is happening in the story.

Step 1

Read the story "The Importance of Trees."

The Importance of Trees

Today is Arbor Day, a day to celebrate the importance of trees. Trees are important for many reasons. They make a gas called oxygen that we need in order to breathe. They are also important because we use trees for making paper. The wood is chopped into little dust-like pieces. Liquid is added to make it into a wet mix called "pulp." The pulp is put through rollers to make paper.

Trees are used in many ways. Trees make a good screen to protect people from strong winds. Trees are also used to build our homes. Even animals use trees to build homes. Beavers build homes from tree branches and mud.

Trees are also important because they grow some of our food. We eat fruits, nuts, and other foods that grow on trees. My favorite red and round fruit—about the size of a baseball—grows on a fairly small tree that can be found all over the United States. This fruit can be eaten raw or used for pies.

Many people plant trees on Arbor Day. My teacher has given us saplings (baby trees) to plant in our yards. The saplings have green needles; they will have cones on them when they grow to maturity and will stay green all year long. My teacher said to plant my tree where it will have sun and room to grow and to give it plenty of water the first year. I can hardly wait for my sapling to grow into a beautiful tree!

© 2005 Englefield & Associates, Inc.   COPYING IS PROHIBITED   35

Activity 7b   *Read on Target* **for Grade 3**

Step 2 — Student Tips

To infer from the story, you need to remember:

- There are clues in the story. Clues are hints the author gives you about the story. Draw a line under each clue, or use your finger to point to the clue.

- What you know will help you figure out the story. Think about what is going on in the story. Have you done it before? Do you know about it?

- Clues, experiences, and knowledge are put together.

- Using clues from the story and what you know will help you figure out what is happening in the story. (This is called an **inference**.)

Step 3 — Complete the reading map. Use the reading map to help you think about inferring.

Activity 7b — *Read on Target* for Grade 3

Map 7b

Infer from the Text

I read clues and use my knowledge to figure out what is happening in the story.

My Clues	+	My Experiences	=	My Inference
Find the clues in each sentence. Write your clues in the boxes below.		Think about your experiences and knowledge of a similar thing. Write your information in the boxes below.		Put the clue and experience/knowledge together to make a guess about what is happening in the story.

Clues to tell you the favorite fruit.

→ **Your experiences and knowledge of a similar thing.**

→ **What is the fruit?**

Clues to tell you the tree.

→ **Your experiences and knowledge of a similar thing.**

→ **What is the tree?**

© 2005 Englefield & Associates, Inc.   COPYING IS PROHIBITED   37

Activity 7b   Read on Target for Grade 3

Step 4
Read the following questions and write your answers.

1. What is the author's favorite fruit?

   _____
   _____
   _____

2. What clues from the story helped you figure out your answer?

   _____
   _____
   _____

3. What kind of tree is the author going to plant on Arbor Day?

   _____
   _____
   _____

4. What clues from the story helped you figure out your answer?

   _____
   _____
   _____

Activity 8a

Predict from the Text

I read clues and use what I know to figure out what will happen in the future.

Step 1   **Read the story "Best in Show."**

Best in Show

Shadow was a good dog. Shadow did whatever Jason asked her to do. Shadow learned to do tricks, and Jason was so proud of her that he wanted other people to see her, too. Jason decided to enter Shadow in the dog show.

Jason told Shadow that to win the "Best-in-Show" prize, dogs have to do many tricks. The dog that wins "Best-in-Show" also has to be a pretty dog.

Jason gave Shadow a bath and brushed her coat. Shadow's hair was clean and shiny. She was a beautiful dog. She was smart, too. She knew lots of hard tricks. She could sit, stay quiet, get the ball, and jump through hoops.

On the day of the dog show, Shadow sat by the side of the ring. Everyone said that Shadow was the most beautiful dog they had ever seen. She waited for her turn to walk around the ring and do her tricks. While she waited, she looked at the other dogs in the dog show.

All of the dogs were sitting quietly. They were also waiting for their turns to walk inside the ring and do some tricks. Soon, all of the dogs had done some tricks. They did not know how to do hard tricks. They only did easy tricks. Then, it was Shadow's turn. She lifted her head and stood beside Jason. It was time to go inside the ring.

© 2005 Englefield & Associates, Inc.

Activity 8a — Read on Target for Grade 3

Step 2 — Student Tips

To predict from the text, you need to remember:

- There are clues in the story. Clues are hints the author gives you about the story. Draw a line under each clue, or use your finger to point to the clue.

- What you know will help you make a guess about what will happen next in the story. Think about what is going on in the story. Have you done it before? Do you know about it?

- Clues, experiences, and knowledge are put together.

- Using clues from the story and what you know will help you figure out what will happen next. (This is called a **prediction**.)

Step 3 — Complete the reading map. Use the reading map to help you think about predicting.

Activity 8a *Read on Target* for Grade 3

Predict from the Text

I read clues and use my knowledge to figure out what will happen next.

Map 8a

My Clues	+	My Experiences	=	My Prediction
Find clues in each sentence or paragraph that would help you answer the questions from the passage.		Think about your experiences and knowledge of a similar thing.		Put the clue and experience/knowledge together to make a guess about what will happen next.

Clues about Shadow listening to Jason. → Your experiences and knowledge of a similar thing. → Predict if Shadow will listen to Jason when she is inside the ring.

Clues about how Shadow looks and what kind of tricks she does. → Your experiences and knowledge of a similar thing. → What do you think will happen when the judges at the dog show give out prizes?

© 2005 ENGLEFIELD & Associates, Inc. COPYING IS PROHIBITED 41

Activity 8a   Read on Target for Grade 3

Step 4
Read the following questions and write your answers.

1. Do you think Shadow will listen to Jason when she is in the ring? Explain your answer using clues from the text or your experiences.

   _____
   _____
   _____

2. What clues from the story helped you figure out if Shadow will listen to Jason?

   _____
   _____
   _____

3. What do you know about a similar thing?

   _____
   _____
   _____

4. Do you predict that Shadow will win the "Best-in-Show" prize? What clues from the story helped you figure out your prediction?

   _____
   _____
   _____

Activity 8b

Predict from the Text

I read clues and use what I know to figure out what will happen in the future.

Step 1 Read the story "The Baseball Game."

The Baseball Game

Joe and Michael got out of the car and ran to the baseball field. Joe is a catcher, and Michael is a pitcher. They have been waiting for this game to start all week, and they have been practicing for it all week. It is the first time they will get to play with their new team.

The coaches called for the team to start the warm-up. Then, the coaches talked about the line-up of players. Everyone was ready to play! Joe went to the catcher's spot, and Michael went to the pitcher's mound.

"This is our big chance to show everyone what we can do, since we practiced our throwing and catching," said Joe.

"That's right. The more we practice, the better we will be. And we practiced a lot!" said Michael. "Look, Joe, there is a searchlight over the trees!"

"That's not a searchlight," said Joe. "That's lightning! It's dangerous to be outside when there's lightning."

Suddenly, the sky became dark with black clouds. The wind started to blow. The beautiful sunny day was changing.

The coaches called out for the team. The coaches needed to talk to the players. Joe and Michael wondered if the coaches would let them stay outside and play baseball.

© 2005 Englefield & Associates, Inc.   COPYING IS PROHIBITED   43

Activity 8b   *Read on Target* for Grade 3

Step 2 — Student Tips

To predict from the text, you need to remember:

- There are clues in the story. Clues are hints the author gives you about the story. Draw a line under each clue, or use your finger to point to the clue.

- What you know will help you make a guess about what will happen next in the story. Think about what is going on in the story. Have you done it before? Do you know about it?

- Clues, experiences, and knowledge are put together.

- Using clues from the story and what you know will help you figure out what will happen next. (This is called a **prediction**.)

Step 3 — Complete the reading map. Use the reading map to help you think about predicting.

Activity 8b   *Read on Target* for Grade 3

Predict from the Text

I read clues and use my knowledge to figure out what will happen next.

Map 8b

My Clues	+	My Experiences	=	My Prediction
Find clues in each sentence or paragraph that would help you answer the questions from the passage.		Think about your experiences and knowledge of a similar thing.		Put the clue and experience/knowledge together to make a guess about what will happen next.

Clues about what the weather will be soon. → Your experiences and knowledge of a similar thing. → What type of weather will happen next?

Clues about whether the team will stay outside. → Your experiences and knowledge of a similar thing. → What will the team do?

Activity 8b　　　　　　　　　　　　Read on Target for Grade 3

Step 4　　Read the following questions and write your answers.

1. What type of weather is going to happen next?

   _____
   _____
   _____

2. What clues from the story helped you predict what type of weather is going to happen?

   _____
   _____
   _____

3. Do you think the team will stay outside and play baseball? Explain your answer.

   _____
   _____
   _____

4. What do you know about staying outside when there is lightning?

   _____
   _____
   _____

Activity 9a

Compare and Contrast

I read to find out how two things are alike and different.

Step 1 — **Read the story "Picture Day."**

Picture Day

Zoë is a basset hound, and Happy Cat is a fluffy cat. They live in the same house. That is just about the only thing they have in common. Zoë's smooth brown and black fur is very different than Happy Cat's white, fluffy fur. Zoë is big and long, and Happy Cat is small and round. Zoë loves to play outdoors. Zoë loves to run and jump. Happy Cat stays inside. Happy Cat sleeps in the warm sun. Happy Cat does not run and jump.

Zoë loves going to new places and meeting new people. Happy Cat is shy and afraid of new places and new people.

It is picture day at the pet store. Samantha will take Zoë and Happy Cat to get their pictures taken at the pet store. They have never been to the pet store before. Zoë barks loudly, while Happy Cat meows softly. When they go into the pet store, they see bright lights. There are many big barking dogs. The whole store smells like a big dog house. Zoë thinks it is a fun place to play. Happy Cat thinks that it is a scary place. Happy Cat just wants to go home.

Zoë loves it! Zoë runs around. Zoë dashes right up to the camera and licks it! Then, Zoë runs to the dog food bag, tears open the bag, and eats it right up!

"Catch that dog!" cries the store manager.

Samantha chases Zoë, and Happy Cat just hides.

Zoë jumps up on the table for her picture. Samantha puts Happy Cat beside her. Zoë licks Happy Cat, and Happy Cat feels safe when Zoë is next to him. They love each other.

Flash! Click! The picture is taken. Zoë and Happy Cat go home.

What an adventure picture day was!

Step 2 — Student Tips

To compare and contrast, you need to remember:

- You are looking at what is special about each thing. Think about what makes something special, like its color and its shape.

- You are checking out how things are alike and different. Compare means to tell how things are alike (the same). Contrast means to tell how things are different (not the same).

Step 3 — Complete the reading map. Use the reading map to help you think about comparing and contrasting.

Activity 9a   *Read on Target* for Grade 3

Map 9a

Compare and Contrast

I read to figure out how two things are alike and different.

Directions: How would you describe the things that you are going to compare and contrast? What shape are they? What color are they? Put a plus sign (+) in a box that matches a characteristic of the other items. Put a minus sign (−) in a box that doesn't match a characteristic of the other items.

Read the names of the things that you are going to compare and contrast in the shaded boxes below.		
Zoë		**Happy Cat**
	+ or −	
Describe the Characteristics. Tell what the things look like. Write your answer in the box next to these characteristics. (Under the shaded boxes.)		
Feelings about each other . . .	+ or −	
Fur	+ or −	
Sounds	+ or −	
Thinks about pet store . . .	+ or −	

© 2005 Englefield & Associates, Inc.   COPYING IS PROHIBITED   49

Step 4: Read the following questions and write your answers.

1. How are Zoë's and Happy Cat's feelings about each other the same?

   _____
   _____
   _____

2. Compare Zoë's fur to Happy Cat's fur.

   _____
   _____
   _____

3. Compare what Zoë and Happy Cat sound like.

   _____
   _____
   _____

4. What do Zoë and Happy Cat think about the pet store?

   _____
   _____
   _____

Activity 9b

Compare and Contrast

I read to find out how two things are alike and different.

Step 1 Read the story "Far Away Favorite."

Far Away Favorite

Mrs. Cooper taught our class about the planets. We studied them one at a time and learned a lot about each one. For our homework, she wants each of us to decide which planet is our favorite. It is really hard for me to choose. I like something about each planet, but I like Mars and Saturn the best.

Mars has interesting facts about it. It has a red color in the night sky. The red color is from large, orange-colored rocks. These rocks make the planet look red from far away. Even though Mars is close to the sun, it's a cold planet. It is even too cold for water. Mars is small when compared to Saturn.

I learned a lot about Saturn that I never knew. Saturn is one of the larger planets. It has famous rings around it, and the rings are made of chunks of ice and dust. Saturn looks yellow in the night sky. The planet is not solid like Mars. Instead, Saturn is made of gases.

Trying to pick my favorite planet is certainly hard. Both are round and can be found in the night sky, but each has neat characteristics of its own. I really like Saturn's rings; however, I think Mars' red color is really pretty. I need to decide quickly—tomorrow is science class!

Activity 9b   *Read on Target* for Grade 3

Step 2

Student Tips

To compare and contrast, you need to remember:

- You are looking at what is special about each thing. Think about what makes something special, like its color and its shape.

- You are checking out how things are alike and different. Compare means to tell how things are alike (the same). Contrast means to tell how things are different (not the same).

Step 3

Complete the reading map. Use the reading map to help you think about comparing and contrasting.

Activity 9b   *Read on Target* for Grade 3

Map 9b

Compare and Contrast

I read to figure out how two things are alike and different.

Directions: How would you describe the things that you are going to compare and contrast? What shape are they? What color are they? Put a plus sign (+) in a box that matches a characteristic of the other items. Put a minus sign (–) in a box that doesn't match a characteristic of the other items.

Read the names of the things that you are going to compare and contrast in the shaded boxes below.	Mars	+ or –	Saturn
Describe the Characteristics. Tell what the things look like. Write your answer in the box next to these characteristics. (Under the shaded boxes.)			
Temperature		+ or –	
Size		+ or –	
Color		+ or –	
What it is made of . . .		+ or –	

© 2005 ENGLEFIELD & ASSOCIATES, Inc.   COPYING IS PROHIBITED   53

Activity 9b   Read on Target for Grade 3

Step 4 — Read the following questions and write your answers.

1. Compare the planets' temperatures. Support your answer with words from the story.

   _____
   _____
   _____

2. Compare or contrast the size of the planets. Support your answer with words from the story.

   _____
   _____
   _____

3. Tell how the planet colors are the same or different. Support your answer with words from the story.

   _____
   _____
   _____

4. Compare or contrast what the planets are made of.

   _____
   _____
   _____

Activity 10a

Analyze the Text by Examining the Use of Fact and Opinion

I figure out if the sentence can be proven or is a personal belief that tells how someone feels or thinks.

Step 1 Read the story "Let's Vote."

Let's Vote

"You have been the best boys and girls since the school year began," said Mrs. Chin. "We are going to vote on how we should spend our free time. Since you're all part of the class, you all get to vote."

"What will our choices be?" asked Jake.

"You will have two choices. Elections always have two choices. One choice will be for extra time at recess. The other choice will be to watch a movie."

The students knew they should all vote. They knew that responsible people vote. They knew that the side with the most votes wins. They knew that they would have a voting day in their classroom. Their voting day would be in November like the real Election Day. Election Day is the Tuesday after the first Monday in November.

The day finally came. It was voting day. Mrs. Chin told the students that each of them should vote. Each child voted and then sat down. Mrs. Chin needed to count the votes to see which choice had the most votes.

"And the winner is . . . more time at recess!" said Mrs. Chin. "There were five people who voted for the movie. There were fifteen people who voted for more time at recess. We will stay out longer at recess on Friday afternoon. Thank you everyone for voting. Everyone should be a responsible person and vote!"

Activity 10a   **Read on Target** for Grade 3

> **Step 2** **Student Tips**

To analyze fact and opinion, you need to remember:

- **Facts** are true for everyone. They can be proven by seeing them or by looking them up.

- **Opinions** are true for some people. Opinions are beliefs about something. Some key words that let you know a sentence is an opinion are best, worst, bad, beautiful, ugly, always, never, everyone, mean, and kind.

- Read more about key words on the Fact and Opinion Worksheet. Then, practice writing fact sentences and opinion sentences.

> **Step 3** **Complete the reading maps. Use the reading maps to help you think about facts and opinions.**

Activity 10a Read on Target for Grade 3

Map 10a.1 — Analyze Aspects of the Text by Examining the Use of Fact and Opinion

I read to figure out if the reading is something that can be proven by evidence or if the reading is a personal belief that tells how someone feels or thinks about something.

Read a sentence from the story.

Everyone should be a responsible person and vote.

← OR →

If you think the sentence is a FACT, place an X in the correct space after the question.	If you think the sentence is an OPINION, place an X in the correct space after the question.
Can the information be seen? Yes _____ No _____ Can you find out if it is true by reading about it? Yes _____ No _____	Is there a Key Word that overstates or describes an opinion? Yes _____ No _____ Does it tell about a belief? Yes _____ No _____
If **one or more** of these answers are yes, then the sentence may be a fact.	If **one or more** of these answers are yes, then the sentence may be an opinion.
AND	AND
This answer **must** be yes for the sentence to be a fact.	This answer **must** be yes for the sentence to be an opinion.
Is the information true for **all** people? Yes _____ No _____	Is the information true for **some** people? Yes _____ No _____

© 2005 ENGLEFIELD & ASSOCIATES, Inc. COPYING IS PROHIBITED

Activity 10a   Read on Target for Grade 3

Map 10a.2 Analyze Aspects of the Text by Examining the Use of Fact and Opinion

I read to figure out if the reading is something that can be proven by evidence or if the reading is a personal belief that tells how someone feels or thinks about something.

Read a sentence from the story.

Election Day is the Tuesday after the first Monday in November.

← OR →

If you think the sentence is a FACT, place an X in the correct space after the question.	If you think the sentence is an OPINION, place an X in the correct space after the question.
Can the information be seen? Yes ____ No ____ Can you find out if it is true by reading about it? Yes ____ No ____	Is there a Key Word that overstates or describes an opinion? Yes ____ No ____ Does it tell about a belief? Yes ____ No ____
If **one or more** of these answers are yes, then the sentence may be a fact.	If **one or more** of these answers are yes, then the sentence may be an opinion.
AND	**AND**
This answer **must** be yes for the sentence to be a fact.	This answer **must** be yes for the sentence to be an opinion.
Is the information true for **all** people? Yes ____ No ____	Is the information true for **some** people? Yes ____ No ____

58   COPYING IS PROHIBITED   © 2005 Englefield & Associates, Inc.

Activity 10a   Read on Target for Grade 3

Map 10a.3   Fact and Opinion Worksheet

Facts	Opinions
• true for everyone • can be proven and supported by evidence and observation • can be checked by looking up the information or seeing it	• true for some people • tells how someone thinks or feels about something • personal belief or judgment about something

You can use key words to change a fact into an opinion. Here are some examples of key words that will help you figure out if the information is an opinion.

KEY WORDS that describe an opinion:
best, great, easy, hard, beautiful, pretty, good, bad, difficult, ugly, terrible, excellent

KEY WORDS that overstate an opinion:
always, never, all, everyone

FACT + KEY WORD = OPINION

Here is an example: 1. This is a book. (fact) 2. This is a great book. (opinion)

Practice changing a fact into an opinion.

Read a fact sentence: **There were five people who voted for a movie.**

Add a KEY WORD to change the fact into an opinion: _____

_____

Practice changing an opinion into a fact.

Read an opinion sentence: **Elections always have two choices.**

Take the KEY WORD out of the opinion sentence and write a fact sentence:

_____

Activity 10a   Read on Target for Grade 3

Step 4
Read the following questions and write your answers.

1. "There were five people who voted for the movie." Is this a fact or an opinion? Explain your answer.

   _____
   _____
   _____

2. The teacher says, "Elections always have two choices." How do you know if this is a fact or an opinion? Explain your answer.

   _____
   _____
   _____

3. Write one factual sentence from the story that explains when election day is.

   _____
   _____
   _____

4. Explain why the sentence "Everyone should be a responsible person and vote" is an opinion.

   _____
   _____
   _____

Activity 10b

Analyze the Text by Examining the Use of Fact and Opinion

I figure out if the sentence can be proven or is a personal belief that tells how someone feels or thinks.

Step 1 Read the story "Bumps or No Bumps."

Bumps or No Bumps

Some of the people on our street want to have the county put speed bumps on our street. They say that too many people go speeding through our neighborhood at 40 miles per hour. My parents say that is a terrible idea. Those people want to make our country neighborhood like a city neighborhood with a lot of rules.

People from the county came and put a meter on our street that counted the number of cars and how fast they were going. We went to a meeting where it was announced that an average of 51 cars went through our neighborhood each day at an average of 32 miles per hour. The speed limit is 30 miles per hour on our street. All people speed, so that didn't sound so bad to me.

At the meeting, the county trustees told the people from our neighborhood that the problem wasn't so bad that speed bumps should be installed. Some kids who play in the street were upset, and so were their parents. All the trustees should care about the danger because every day there are kids riding their bikes on the street, and these kids could be injured by speeding cars.

Other people on our street said that speed bumps can damage cars that have to drive over them day after day. In the end, the trustees said they would change the speed limit to 25 miles per hour.

Not everyone was happy, but the 28 people who came to the meeting were glad that they were allowed to give their opinions.

© 2005 Englefield & Associates, Inc.   COPYING IS PROHIBITED

Activity 10b   Read on Target for Grade 3

Step 2 — Student Tips

To analyze fact and opinion, you need to remember:

- **Facts** are true for everyone. They can be proven by seeing them or by looking them up.

- **Opinions** are true for some people. Opinions are beliefs about something. Some key words that let you know a sentence is an opinion are best, worst, bad, beautiful, ugly, always, never, everyone, mean, and kind.

- Read more about key words on the Fact and Opinion Worksheet. Then, practice writing fact sentences and opinion sentences.

Step 3 — Complete the reading maps. Use the reading maps to help you think about facts and opinions.

Activity 10b *Read on Target* for Grade 3

Map 10b.1 — Analyze Aspects of the Text by Examining the Use of Fact and Opinion

I read to figure out if the reading is something that can be proven by evidence or if the reading is a personal belief that tells how someone feels or thinks about something.

Read a sentence from the story.

The speed limit is 30 miles per hour on our street.

← OR →

If you think the sentence is a FACT, place an X in the correct space after the question.	If you think the sentence is an OPINION, place an X in the correct space after the question.
Can the information be seen? Yes ____ No ____ Can you find out if it is true by reading about it? Yes ____ No ____	Is there a Key Word that overstates or describes an opinion? Yes ____ No ____ Does it tell about a belief? Yes ____ No ____
If **one or more** of these answers are yes, then the sentence may be a fact.	If **one or more** of these answers are yes, then the sentence may be an opinion.
AND	AND
This answer **must** be yes for the sentence to be a fact.	This answer **must** be yes for the sentence to be an opinion.
Is the information true for **all** people? Yes ____ No ____	Is the information true for **some** people? Yes ____ No ____

Activity 10b   Read on Target for Grade 3

Map 10b.2

Analyze Aspects of the Text by Examining the Use of Fact and Opinion

I read to figure out if the reading is something that can be proven by evidence or if the reading is a personal belief that tells how someone feels or thinks about something.

Read a sentence from the story.

All the trustees should care about the danger because every day there are kids riding their bikes in the street who could be injured by speeding cars.

← **O R** →

If you think the sentence is a FACT, place an X in the correct space after the question.	If you think the sentence is an OPINION, place an X in the correct space after the question.
Can the information be seen? Yes ____ No ____ Can you find out if it is true by reading about it? Yes ____ No ____	Is there a Key Word that overstates or describes an opinion? Yes ____ No ____ Does it tell about a belief? Yes ____ No ____
If **one or more** of these answers are yes, then the sentence may be a fact.	If **one or more** of these answers are yes, then the sentence may be an opinion.
AND	AND
This answer **must** be yes for the sentence to be a fact.	This answer **must** be yes for the sentence to be an opinion.
Is the information true for **all** people? Yes ____ No ____	Is the information true for **some** people? Yes ____ No ____

Activity 10b   *Read on Target* for Grade 3

Map 10b.3 — Fact and Opinion Worksheet

Facts	Opinions
• true for everyone • can be proven and supported by evidence and observation • can be checked by looking up the information or seeing it	• true for some people • tells how someone thinks or feels about something • personal belief or judgment about something

You can use key words to change a fact into an opinion. Here are some examples of key words that will help you figure out if the information is an opinion.

KEY WORDS that describe an opinion:
best, great, easy, hard, beautiful, pretty, good, bad, difficult, ugly, terrible, excellent

KEY WORDS that overstate an opinion:
always, never, all, everyone

FACT + KEY WORD = OPINION

Here is an example: 1. This is a book. (fact) 2. This is a great book. (opinion)

Practice changing a fact into an opinion.

Read a fact sentence: **The speed limit is 30 miles per hour.**

Add a KEY WORD to change the fact into an opinion: _____

_____

Practice changing an opinion into a fact.

Read an opinion sentence: **All people speed, so that didn't sound bad to me.**

Take the KEY WORD out of the opinion sentence and write a fact sentence:

_____

© 2005 Englefield & Associates, Inc.   COPYING IS PROHIBITED

Activity 10b   **Read on Target** for Grade 3

Step 4
Read the following questions and write your answers.

1. All the trustees should care about the danger because every day there are kids riding their bikes on the street, and these kids could be injured by speeding cars. Is this a fact or an opinion? Explain your answer.

   _____
   _____
   _____

2. "The speed limit is 30 miles per hour." Where could you find information to prove whether this is a fact or an opinion?

   _____
   _____
   _____

3. Write one factual sentence from the story that tells how fast cars travel on the street.

   _____
   _____
   _____

4. Explain why the sentence "All people speed, so that didn't sound so bad to me" is an opinion.

   _____
   _____
   _____

Activity 11a

Explain How and Why an Author Uses Contents of a Text to Support His/Her Purpose for Writing

I tell the process (how) and the reason (why) the story was written.

Step 1 Read the selection "Eat Right and Be Healthy."

Eat Right and Be Healthy

Scientists are worried about the health of the American people. They think that some adults do not eat the right foods. They also think some adults do not get enough exercise. As people grow older, poor eating habits and little exercise will lead them into poor health. They may need to go to the doctor's office or the hospital more often.

There is an answer to this problem, and you can help. You can look more closely at the foods you eat. Eat more healthy foods. Get rid of the chips and the junk food, and stop eating at fast food restaurants. If you go to restaurants, make healthier choices. You could ask for a salad instead of macaroni and cheese or have applesauce instead of french fries.

Eat fruits and vegetables with all meals. Have fruits and vegetables for snacks during the day. After school, have a fruit or vegetable for a snack. Do not snack before bed. These are steps you can do now as a child so you will stay healthy as you grow into adulthood.

Exercising is another way to stay healthy. Exercise more often. Turn off the television, and get off the couch. Run and play outside. Exercise for 30 to 90 minutes every day. Walk around the neighborhood, or ride your bike.

If you learn better habits as a child, you will grow into a healthy adult. If you start soon enough, you will learn how to eat right and stay healthy. You will stay out of the doctor's office and the hospital. Then, scientists can spend time fighting diseases instead of helping the American people learn to eat right and exercise.

Activity 11a   *Read on Target* **for Grade 3**

Step 2 — Student Tips

To explain the author's purpose for writing, you need to remember:

- Tell why. "Why" is the reason the author wrote the text.
- Tell how. Find words or pictures that help you figure out the author's writing process.
- Look for clues that tell you why authors write:

  Enjoyment (Funny sentences, interesting words, and images)

  Understand (Words that tell you what people (characters) are like and what they think, feel, or do)

  Find Out/Learn (Facts, charts, graphs, and pictures)

  Solve Problems (Words that tell you about an action or a decision)

  Persuade (Words that tell you how you should think)

Step 3 — Complete the reading map. Use the reading map to help you think about the author's purpose.

Activity 11a   Read on Target for Grade 3

| Map 11a | Explain How and Why an Author Uses Contents of a Text to Support His/Her Purpose for Writing |

I tell the process (how) and the reason (why) the story was written.

Why Authors Write (Purpose)

 Enjoyment

 Solve a problem

 Persuade you to agree

Understand about what people (characters) think, feel, or do.

Find out or learn about something

Circle the type of writing.	Draw a picture or write one reason why the text was written.
fiction poetry nonfiction	

Write the sentences from the text that show how the author tells the purpose for writing.

© 2005 Englefield & Associates, Inc.   COPYING IS PROHIBITED   69

Activity 11a   Read on Target for Grade 3

Step 4
Read the following questions and write your answers.

1. What is the type of writing?

   _____

   _____

   _____

2. Why was the text written?

   _____

   _____

   _____

3. Explain how you figured out why the text was written.

   _____

   _____

   _____

4. How does the author show his or her purpose for writing?

   _____

   _____

   _____

Activity 11b

Explain How and Why an Author Uses Contents of a Text to Support His/Her Purpose for Writing

I tell the process (how) and the reason (why) the story was written.

Step 1 Read the story "How to Train a Cat."

How to Train a Cat

Most people think that you can't train a cat. They may be right. Training your cat may really be training yourself. If your cat doesn't want to do something, it won't. But you can adjust your behavior to help the cat do what you want.

My cat Frisky liked to wake me up at night. If I ignored him, he would claw at the bed skirt and the other furniture. That would get me up to yell at him not to claw the furniture, so that is what he did to get me out of bed and give him food. I realized Frisky was training me in a bad way.

One night, I thought, "This has to stop." When he came in and clawed on the furniture, I tossed a pillow in his direction and said, "No, Frisky," in a very stern voice, but I didn't get up. I said, "Say meow," several times and meowed like a cat. Finally, he meowed back, and I got up and gave him food and praise. This happened a few more times. Now, Frisky meows to get me up instead of clawing the furniture.

I haven't figured out how to train him not to wake me up in the middle of the night, though.

Step 2: Student Tips

To explain the author's purpose for writing, you need to remember:

- Tell why. "Why" is the reason the author wrote the text.
- Tell how. Find words or pictures that help you figure out the author's writing process.
- Look for clues that tell you why authors write:

  Enjoyment (Funny sentences, interesting words, and images)

  Understand (Words that tell you what people (characters) are like and what they think, feel, or do)

  Find Out/Learn (Facts, charts, graphs, and pictures)

  Solve Problems (Words that tell you about an action or a decision)

  Persuade (Words that tell you how you should think)

Step 3: Complete the reading map. Use the reading map to help you think about the author's purpose.

Activity 11b   Read on Target for Grad 3

Map 11b — Explain How and Why an Author Uses Contents of a Text to Support His/Her Purpose for Writing

I tell the process (how) and the reason (why) the story was written.

Why Authors Write (Purpose)

- Enjoyment
- Solve a problem
- Persuade you to agree
- Understand about what people (characters) think, feel, or do.
- Find out or learn about something

Circle the type of writing.	Draw a picture or write one reason why the text was written.
fiction poetry nonfiction	

Write the sentences from the text that show how the author tells the purpose for writing.

© 2005 Englefield & Associates, Inc.   COPYING IS PROHIBITED

Activity 11b   Read on Target for Grade 3

Step 4 Read the following questions and write your answers.

1. What is the type of writing?

   _____
   _____
   _____

2. Why was the text written?

   _____
   _____
   _____

3. Explain how you figured out why the text was written.

   _____
   _____
   _____

4. Write a sentence from the text that tells you how the author shows his or her purpose for writing.

   _____
   _____
   _____

Activity 12a

Identify Main Idea/Supporting Details

I figure out the overall idea and supporting details of the story.

Step 1 Read the story "Deserts."

Deserts

Deserts are dry land with a lot of sand and rocks. The sun shines most of the day, and it is hot. Very little rain falls in a desert. The sand and rocks are everywhere.

Only animals and plants that can survive the dry climate can be found in the desert. Cactuses are plants that can be found in the desert. You may see a fox. Sometimes, a little mouse might be found running around in the desert.

Activity 12a   *Read on Target* for Grade 3

Step 2 — Student Tips

To identify the main idea and supporting details, you need to remember:

- There is a main idea sentence in each paragraph. The main idea sentence tells what the paragraph is about.

- There are detail sentences in each paragraph. Detail sentences support the main idea, explain the main idea, and give information about the main idea.

Step 3

Complete the reading map. Use the reading map to help you think about the main idea and supporting details.

Activity 12a

Read on Target **for Grade 3**

Map 12a — Identify Main Idea/Supporting Details
I figure out the overall idea and supporting details of the story.

Read each paragraph to tell what the selection is about.

What is the main idea of paragraph 1?

_____
_____
_____

What is the main idea of paragraph 2?

_____
_____
_____

Details: Write words from the selection that support, explain, or give information about the main idea of paragraph 1.

_____
_____
_____
_____

Details: Write words from the selection that support, explain, or give information about the main idea of paragraph 2.

_____
_____
_____
_____

When I think about the main ideas of paragraphs 1 and 2, I decide the **overall** main idea of the selection is . . .

_____
_____
_____

© 2005 Englefield & Associates, Inc.   COPYING IS PROHIBITED

Activity 12a Read on Target for Grade 3

Step 4
Read the following questions and write your answers.

1. What is the main idea of the first paragraph?

   _____
   _____
   _____

2. Give one detail from the selection that supports the main idea of paragraph 1.

   _____
   _____

3. What is the main idea of the second paragraph?

   _____
   _____
   _____

4. Give one detail from the selection that supports the main idea of paragraph 2.

   _____
   _____

5. In your own words, write the main idea of the whole selection.

   _____
   _____
   _____

Activity 12b

Identify Main Idea/Supporting Details

I figure out the overall idea and supporting details of the story.

Step 1 Read the story "Grooming Your Cat."

Grooming Your Cat

Grooming your cat can be good both for your cat and for you. The veterinarian told me grooming would help my cat, Whiskers. He would not get so many hairballs (balls of hair he swallowed while licking his fur). He said holding and brushing my cat would help him relax.

This is how I helped my cat love his grooming. First, I let him smell the brush. He smelled it with his nose and touched it with his paw. Next, I stroked his back with it. Then, I held him and brushed him for about five minutes. I brushed him almost every day for about a week. Now, if I want Whiskers to come to me, I just get out his brush, and he comes to get his grooming.

© 2005 Englefield & Associates, Inc. COPYING IS PROHIBITED

Activity 12b   *Read on Target* for Grade 3

Step 2 **Student Tips**

To identify the main idea and supporting details, you need to remember:

- There is a main idea sentence in each paragraph. The main idea sentence tells what the paragraph is about.

- There are detail sentences in each paragraph. Detail sentences support the main idea, explain the main idea, and give information about the main idea.

Step 3 **Complete the reading map. Use the reading map to help you think about the main idea and supporting details.**

Activity 12b Read on Target for Grade 3

Map 12b — Identify Main Idea/Supporting Details
I figure out the overall idea and supporting details of the story.

Read each paragraph to tell what the selection is about.

What is the main idea of paragraph 1?

_____
_____
_____

What is the main idea of paragraph 2?

_____
_____
_____

Details: Write words from the selection that support, explain, or give information about the main idea of paragraph 1.

_____
_____
_____
_____

Details: Write words from the selection that support, explain, or give information about the main idea of paragraph 2.

_____
_____
_____
_____

When I think about the main ideas of paragraphs 1 and 2, I decide the **overall** main idea of the selection is . . .

_____
_____
_____

© 2005 ENGLEFIELD & ASSOCIATES, Inc.   COPYING IS PROHIBITED

Activity 12b   Read on Target for Grade 3

Step 4

Read the following questions and write your answers.

1. What is the main idea of the first paragraph?

   _____

   _____

   _____

2. Give one detail from the selection that supports the main idea of paragraph 1.

   _____

   _____

3. What is the main idea of the second paragraph?

   _____

   _____

   _____

4. Give one detail from the selection that supports the main idea of paragraph 2.

   _____

   _____

5. In your own words, write the main idea of the whole selection.

   _____

   _____

   _____

COPYING IS PROHIBITED   © 2005 Englefield & Associates, Inc.

Activity 13a

Respond to the Text
I tell my thoughts, feelings, or observations of something similar to what happens in the story.

Step 1   Read the selection "Building New Houses or Destroying Old Homes?"

Building New Houses or Destroying Old Homes?

Our town is going to change a lot of land into places for new houses. Builders plan to build 100 new homes. I heard there would be a small park, too. It is exciting to be part of a growing town. However, I want to know what will happen to the plants and animals who used to call this land their home.

My teacher taught us about animal habitats like forests and ponds. We learned that many different animals live there. There are woods where the new houses are going to go. We saw all kinds of animals when we hiked in the woods. The trees had insects in the bark and animals in the branches. Our teacher taught us that dead trees make great homes for animals like raccoons and bats. It makes me sad to think that these animals' homes might be taken away.

Animals living in the woods might lose their homes, and so will the animals living in the pond and the fields. Some of these animals, like frogs, need both water and dry land to live. Building homes will hurt insects living on and around water. The birds that eat the pond insects will have less food. I worry that these animal changes will hurt the food chain. Animal habitats will change and may even disappear from this area.

On the other hand, building these houses will make more jobs. Builders will pay the workers for building the houses. More people living here will mean more money spent at the stores. The town might even need to build more stores, which would give even more people jobs.

It is exciting to see towns grow, but it is scary, too. I wonder what will happen to the plants and animals. I am excited to see building, especially when it gives people more chances for jobs and more places to live. But when we build new houses, we also destroy some old homes.

Step 2 — Student Tips

To respond to the text, you need to remember:

- Tell your thoughts, feelings, and things you do or see that are like things in the story. Does the character think or feel like you do? Does the character act or talk like you? Does the setting make you think about something that has happened to you?

- Look for feeling words to describe emotions. Some emotion words like happy, sad, and afraid are on the reading map. Use these words to tell how you feel about the story.

Step 3 — Complete the reading map. Use the reading map to help you think about the story.

Activity 13a   Read on Target for Grade 3

Map 13a

Respond to the Text
I tell my thoughts, feelings, experiences, or observations of something similar to what happens in the story.

After reading the story, tell your thoughts and feelings of something similar that happened to you. Use your own feeling words or use the bank below to help you.

_____

_____

_____

After reading the story, tell about experiences or observations you have had of a similar thing.

_____

_____

_____

Bank of words to describe how you might feel

Happy	Angry	Afraid	Depressed
Glad	Enraged	Frightened	Down
Pleased	Furious	Scared	Rotten
Proud	Mad	Shaky	Sad
Wonderful	Upset	Worried	Tearful

© 2005 Englefield & Associates, Inc.   COPYING IS PROHIBITED

Activity 13a   Read on Target for Grade 3

Step 4
Read the following questions and write your answers.

1. Tell two changes that are planned for the author's town.

_____
_____
_____

2. How does the author feel about taking away animals' homes?

_____
_____
_____

3. The author discusses hurting the animal food chain. How does that make you feel?

_____
_____
_____

4. Some people are excited about building houses. Tell about a time when you felt excited.

_____
_____
_____

Activity 13b

Respond to the Text
I tell my thoughts, feelings, or observations of something similar to what happens in the story.

Step 1 Read the story "Science Careers."

Science Careers

Students think about what they will be when they grow up. Now is a good time for young people to try a science career. Science is exciting and fun. Science shows the past. Science can be about right here and right now. Science also discovers the future. Examples of science careers include a science teacher, a biologist, or a nature guide.

Science teachers teach all types of science. They teach about plants and animals. They teach about light and sound movement. Children get excited when they learn about space and the planets. Some students even get to learn about dinosaurs and how they became extinct.

After learning about many areas of science, some children decide that they really like plants and animals. The career that they may choose is a biologist. Biologists enjoy studying about plant and animal life. They try to understand how plants and animals behave.

Information about plants and animals sometimes leads children to want to become nature guides. These people study the plants and animals in different environments. They become angry and upset when they see how pollution hurts plants and animals. Sometimes they become scared about what will happen to the plants and animals living in polluted areas. Nature guides are one of many careers in science.

There are many ways studying science helps people. When teachers provide lessons in science class, children get excited about all the possibilities of science in their lives. Some students may even make the choice to have a career in science.

Activity 13b   **Read on Target** for Grade 3

Step 2 Student Tips

To respond to the text, you need to remember:

- Tell your thoughts, feelings, and things you do or see that are like things in the story. Does the character think or feel like you do? Does the character act or talk like you? Does the setting make you think about something that has happened to you?

- Look for feeling words to describe emotions. Some emotion words like happy, sad, and afraid are on the reading map. Use these words to tell how you feel about the story.

Step 3 Complete the reading map. Use the reading map to help you think about the story.

Activity 13b   Read on Target for Grade 3

Map 13b

Respond to the Text
I tell my thoughts, feelings, experiences, or observations of something similar to what happens in the story.

After reading the story, tell your thoughts and feelings of something similar that happened to you. Use your own feeling words or use the bank below to help you.

_____

_____

_____

After reading the story, tell about experiences or observations you have had of a similar thing.

_____

_____

_____

Bank of words to describe how you might feel

Happy	Angry	Afraid	Depressed
Glad	Enraged	Frightened	Down
Pleased	Furious	Scared	Rotten
Proud	Mad	Shaky	Sad
Wonderful	Upset	Worried	Tearful

© 2005 Englefield & Associates, Inc.   COPYING IS PROHIBITED   89

Activity 13b   Read on Target for Grade 3

Step 4

Read the following questions and write your answers.

1. How do some students feel when they learn about space and planets?

   _____

   _____

   _____

2. Biologists enjoy studying about plants and animals. Do you enjoy studying about plants and animals? Explain your answer.

   _____

   _____

   _____

3. Why do nature guides sometimes become angry and upset?

   _____

   _____

   _____

4. The text states that sometimes nature guides become frightened about plants and animals living in polluted areas. How do you feel about plants and animals living in polluted areas?

   _____

   _____

Activity 14a

Evaluate and Critique the Text

I tell about the strengths and the weaknesses of what I read.

Step 1

Read the story "Keep or Release."

Keep or Release

Fishing is my favorite hobby. My family and I go fishing a lot in the summer. We take our boat to a lake where we camp. We put the boat in the water. Then, we turn on the motor and steer the boat out to where we think we will catch some fish. We have a special machine on our boat called a fish finder. You turn it on, and you can see the shapes of fish on the screen. We really enjoy being out in the fresh air on the lake. We talk and fish for hours.

Sometimes we catch a lot of fish. Then we scale them. That means to scrape off the scales. Next, my mom and dad filet them; that means to cut the meat off the sides of the fish. Then we cook them over the campfire, and they taste great.

Last week, I brought my friend Bobby with me. He seemed excited at first to go camping and fishing with us, but then when we got out on the lake, he didn't want to fish. I thought he didn't want to bait the hook. I told him I would put the worm on his hook for him. He said that he really didn't want to fish but he would watch me.

He was quiet for a while. Then he told me that his father fished, too. "My dad lets all his fish go. He releases the little ones so that they could grow up and become bigger fish. He releases the big ones so that they can breed more fish. My dad has fun fishing, but the fish can still live."

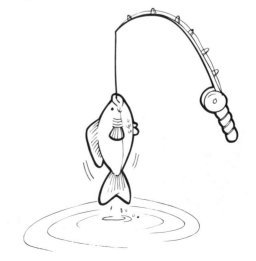

© 2005 Englefield & Associates, Inc.   COPYING IS PROHIBITED   91

Activity 14a   Read on Target for Grade 3

My dad looked at Bobby and said, "Bobby, we throw back the fish that are too small or very large, but we like to eat what we catch. We feel that if you are going to fish, you should eat what you catch and feed your family. Otherwise, you are just having fun and stressing the fish."

Bobby looked at my dad and said, "Mr. Snyder, I never thought of it that way. I always thought you should catch and release the fish when you are fishing. We don't really eat fish at home."

"Well, Bobby," my dad said, "it is just another way of looking at it."

Step 2 — Student Tips

To evaluate and critique the text, you need to remember:

- List the strengths (good points) and weaknesses (bad points).

- See if the information agrees with the question. Ask yourself, "Does this information agree with the question? Does this information not agree with the question?"

- Tell why you think the answer is right.

Step 3

Complete the reading map. Use the reading map to help you think about how to evaluate and critique the text.

Activity 14a Read on Target for Grade 3

Map 14a

Evaluate and Critique the Text
I tell about the strengths and the weaknesses of what I read.

Did the author make you think keeping the fish you catch is a good idea? Why?

The strengths are...	The weaknesses are...
_____	_____
_____	_____
_____	_____
_____	_____
_____	_____
_____	_____
_____	_____

After reading the strengths and weaknesses, the answer to the question is:
_____
_____
_____
_____

Activity 14a Read on Target for Grade 3

Step 4 — Read the following questions and write your answers.

1. Write one reason that keeping and eating the fish you catch is good.

   _____
   _____
   _____

2. Write another reason why fishing is good.

   _____
   _____
   _____

3. Write one reason why Bobby doesn't think you should keep the fish you catch.

   _____
   _____
   _____

4. Do you think that keeping the fish you catch to eat is a good idea or a bad idea?

   _____
   _____
   _____

Activity 14b

Evaluate and Critique the Text

I tell about the strengths and the weaknesses of what I read.

Step 1   **Read the story "Abuela's Gift."**

Abuela's Gift

When Beatriz was eight years old, her grandmother came to live with her family. Beatriz always called her grandmother "Abuela," the Spanish word for "grandmother." Abuela brought gifts from Mexico to each member of the family. Everyone was excited to open his or her present. Then, Abuela gave everyone the most delicious chocolate candy and fruit. Beatriz told her mother how much she liked Abuela's gifts. Her mother smiled and said that all of the gifts were very nice, but having Abuela come to live with them was the best gift of all. Beatriz could not understand how a person could be a present, but she was happy that Abuela was just down the hall.

Soon, Abuela's voice was the waking sound Beatriz heard each morning. The songs always sounded so cheerful and sweet. Abuela could quickly braid Beatriz's long black hair without pulling or hurting her head. Abuela was never too busy to listen to the things Beatriz wanted to share. Abuela learned to play checkers and crazy eights. She told funny stories about when Beatriz's father was a little boy. Best of all, Abuela seemed to always have a hug or a smile for everyone. In no time at all, Beatriz began to understand what her mother had meant. Abuela's gifts from Mexico had been nice to receive, but having Abuela there every day was the best gift.

Activity 14b **Read on Target** for Grade 3

Step 2 — Student Tips

To evaluate and critique the text, you need to remember:

- List the strengths (good points) and weaknesses (bad points).
- See if the information agrees with the question. Ask yourself, "Does this information agree with the question? Does this information not agree with the question?"
- Tell why you think the answer is right.

Step 3

Complete the reading map. Use the reading map to help you think about how to evaluate and critique the text.

Activity 14b

Read on Target for Grade 3

Map 14b

Evaluate and Critique the Text
I tell about the strengths and the weaknesses of what I read.

Did the author do a good job of explaining how a person can be a gift?

The strengths are...	The weaknesses are...

After reading the strengths and weaknesses, the answer to the question is:

Activity 14b   Read on Target for Grade 3

Step 4 — Read the following questions and write your answers.

1. Write two gifts that Abuela gave to Beatriz.

   _____
   _____
   _____

2. Write two ways Abuela was like a gift to the family.

   _____
   _____
   _____

3. What weakness or missing information do you wish the author had included in this story?

   _____
   _____
   _____

4. What were the author's strengths in showing how Abuela could be a gift? Explain your answer.

   _____
   _____
   _____

Activity 15a

Summarize the Text

I tell the overall meaning of the text in my own words.

Step 1   Read the story "The Heart."

The Heart

Your heart is a very important muscle that is small but mighty. It is about the size of your fist. It weighs less than one pound. The heart is divided into two sides. Each side is divided again into top and bottom sections. Altogether, there are four sections of the heart. The heart muscle works all the time, whether you're running, playing, or just relaxing.

The heart pumps your blood. Your heart pumps blood to the lungs. The heart also pumps blood from the lungs to the rest of your body to feed cells. If you run around for a while, your heart pumps the blood faster than before. When your heart is at rest, it only beats 60–80 times a minute. When you exercise, your heart beats up to 200 times in a minute. It's the most amazing muscle in your body.

Activity 15a Read on Target for Grade 3

Step 2 — Student Tips

To summarize the text, you need to remember:

- Each paragraph has a main idea. Tell only the important information in each paragraph. Leave out unimportant information. Use your own words (different words that mean the same thing as the words in the story). Make sure you stay on the topic.

- You should follow these steps.

    1. Read the whole text.
    2. Next, read one paragraph at a time.
    3. Circle important information. This will help you know what to write on your reading map.
    4. Write one sentence that tells the main idea of each paragraph in your own words.
    5. Last, write the overall idea of the whole text in your own words.

Step 3 — Complete the reading maps. Use the reading maps to help you think about your summary.

Activity 15a — Read on Target for Grade 3

Map 15a.1

Summarize the Text
I tell the overall meaning of what I read in my own words.

Summary Sentences
Cross out the unimportant information from paragraph 1.
Circle the most important details.

> Your heart is a very important muscle that is small but mighty. It is about the size of your fist. It weighs less than one pound. The heart is divided into two sides. Each side is divided again into top and bottom sections. Altogether, there are four sections of the heart. The heart muscle works all the time, whether you're running, playing, or just relaxing.

Think about the most important information that you circled. Now, use your own words to write one sentence that tells the overall idea of the paragraph.

___

___

___

___

© 2005 Englefield & Associates, Inc.   COPYING IS PROHIBITED

Activity 15a   Read on Target for Grade 3

Map 15a.2

Summarize the Text
I tell the overall meaning of what I read in my own words.

Summary Sentences
Cross out the unimportant information from paragraph 2.
Circle the most important details.

The heart pumps your blood. Your heart pumps blood to the lungs. The heart also pumps blood from the lungs to the rest of your body to feed cells. If you run around for a while, your heart pumps the blood faster than before. When your heart is at rest, it only beats 60–80 times a minute. When you exercise, your heart beats up to 200 times in a minute. It's the most amazing muscle in your body.

Think about the most important information that you circled. Now, use your own words to write one sentence that tells the overall idea of the paragraph.

_____
_____
_____
_____

Activity 15a

Read on Target for Grade 3

Map 15a.3

Summarize the Text
I tell the overall meaning of what I read
in my own words.

Circle the sentence that summarizes the whole story.

You need your heart muscle to pump your
blood and keep you alive.

Your heart is inside your body and under your ribs.

Activity 15a *Read on Target* **for Grade 3**

Step 4
Read the following questions and write your answers.

1. Write one detail from paragraph 1 that tells about the size of your heart.

   _____
   _____
   _____

2. Write a summary sentence for paragraph 1.

   _____
   _____
   _____

3. Write a summary sentence for paragraph 2 that tells what the heart does that is important.

   _____
   _____
   _____

4. Write a sentence to summarize the entire text.

   _____
   _____
   _____

Activity 15b

Summarize the Text

I tell the overall meaning of the text in my own words.

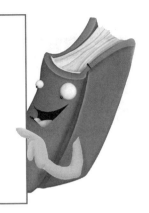

Step 1

Read the story "The Life Cycle of a Frog."

The Life Cycle of a Frog.

The frog's life cycle works well with its habitat (where it lives). The frog is in the class of animals called amphibians. Amphibians spend some of their lives on land and some in the water. To live in both places, the frog changes as it grows. To begin the life cycle, the female frog lays her eggs in the water. The eggs need water to keep them wet. Soon, the eggs hatch into tadpoles. Tadpoles live only in the water. A tadpole has a tail to help it swim through the water. Tadpoles look like small fish with big bodies.

As tadpoles get older, they do not look as much like fish anymore. Their tails begin to shrink. Legs start to appear. The tadpoles start to look more like frogs as their legs grow. Finally, the tadpole loses its tail.

An adult frog has webbed feet that help it swim in the water, but it also has four legs that help it move around on land. Frogs can easily hop wherever they want to go. The frog's life cycle lets the frog live both in and out of the water.

Activity 15b

Read on Target for Grade 3

Step 2 — Student Tips

To summarize the text, you need to remember:

- Each paragraph has a main idea. Tell only the important information in each paragraph. Leave out unimportant information. Use your own words (different words that mean the same thing as the words in the story). Make sure you stay on the topic.

- You should follow these steps.

    1. Read the whole text.
    2. Next, read one paragraph at a time.
    3. Circle important information. This will help you know what to write on your reading map.
    4. Write one sentence that tells the main idea of each paragraph in your own words.
    5. Last, write the overall idea of the whole text in your own words.

Step 3 — Complete the reading maps. Use the reading maps to help you think about your summary.

Activity 15b   Read on Target for Grade 3

Map 15b.1

Summarize the Text
I tell the overall meaning of what I read in my own words.

Summary Sentences
Cross out the unimportant information from paragraph 1.
Circle the most important details.

> The frog's life cycle works well with its habitat (where it lives). The frog is in the class of animals called amphibians. Amphibians spend some of their lives on land and some in the water. To live in both places, the frog changes as it grows. To begin the life cycle, the female frog lays her eggs in the water. The eggs need water to keep them wet. Soon, the eggs hatch into tadpoles. Tadpoles live only in the water. A tadpole has a tail to help it swim through the water. Tadpoles look like small fish with big bodies.

Think about the most important information that you circled. Now, use your own words to write one sentence that tells the overall idea of the paragraph.

_____

_____

_____

_____

Activity 15b Read on Target for Grade 3

Map 15b.2

Summarize the Text
I tell the overall meaning of what I read in my own words.

Summary Sentences
Cross out the unimportant information from paragraph 2. Circle the most important details.

As tadpoles get older, they do not look as much like fish anymore. Their tails begin to shrink. Legs start to appear. The tadpoles start to look more like frogs as their legs grow. Finally, the tadpole loses its tail.

Think about the most important information that you circled. Now, use your own words to write one sentence that tells the overall idea of the paragraph.

_____
_____
_____
_____

Activity 15b Read on Target for Grade 3

Map 15b.3

Summarize the Text
I tell the overall meaning of what I read in my own words.

Summary Sentences
Cross out the unimportant information from paragraph 3.
Circle the most important details.

An adult frog has webbed feet that help it swim in the water, but it also has four legs that help it move around on land. Frogs can easily hop wherever they want to go. The frog's life cycle lets the frog live both in and out of the water.

↓ ↓

Think about the most important information that you circled. Now, use your own words to write one sentence that tells the overall idea of the paragraph.

_____
_____
_____
_____

Activity 15b　　　　　　　　　　　　　　*Read on Target* for Grade 3

Map 15b.4　　　Summarize the Text
I tell the overall meaning of what I read
in my own words.

Circle the sentence that summarizes the whole story.

Frogs and tadpoles are amphibians.

As a frog grows, its body changes to be able to live in and out of water.

Activity 15b *Read on Target* for Grade 3

Step 4
Read the following questions and write your answers.

1. What detail from paragraph 1 tells where tadpoles live?

2. Write one summary sentence for paragraph 2 that tells what happens to a tadpole as it gets older.

3. Write a detail sentence about how adult frogs move around.

4. Write a sentence to summarize the entire text.

Activity 16a

Identify Cause and Effect

I read to find out the reason why something happened and what happened.

Step 1 Read the story "The Smallest Rabbit."

The Smallest Rabbit

One day, Mrs. Jones looked outside of her house and saw a basket on her front step. It was a very big basket with something in it that was moving. She went outside to see what was inside the basket. There were three baby rabbits inside. One rabbit was spotted and brown, one was black with white feet, and one rabbit was all white. The brown and black rabbits were bigger than the white rabbit and very active. They hopped about most of the time that they weren't asleep. The little white rabbit was the smallest in the basket. It was fluffy and very calm. All of the rabbits were very hungry when Mrs. Jones found them.

Mrs. Jones knew that she could not keep the three rabbits, so she told her friends and neighbors about the rabbits and put a sign up in her yard. The sign said "FREE RABBITS." A lot of people came to see the rabbits. One person chose the black rabbit with white feet to take home with him. Another chose the brown rabbit, but no one seemed to want the little white fluffy rabbit.

Mrs. Jones was very sad. She wanted the rabbits to have nice homes. She wanted someone to love the small, white, fluffy rabbit. She put a new sign in her yard and hoped someone would stop by.

A few days later, a little girl was riding her bike past Mrs. Jones's house. Her name was Olivia. Olivia saw the sign in the family's yard that said "ONE FREE FLUFFY WHITE RABBIT." She looked inside the basket. Olivia loved animals. She loved anything fluffy and soft. She thought that she would love this rabbit.

The little rabbit heard Mrs. Jones say, "Olivia, would you like to take this rabbit home?"

Olivia paused to look more closely at the rabbit. "Oh, look! What a small rabbit!" said Olivia. "I would really like to have it."

The small, fluffy, white rabbit looked out of the basket. This girl was talking about him! The rabbit looked at the small girl with big blue eyes who was looking back at him. "She's tiny, too," thought the small rabbit. "Just like me!"

Activity 16a   Read on Target for Grade 3

Step 2 — Student Tips

To identify cause and effect, you need to remember:

- The **cause** is the action or the reason that makes something happen (why something happened).
- The **effect** is the result (what happened).
- Look for key words that are clues to help you figure out if the statement is a cause or an effect.
    - Some of the key words that tell you why something happened (cause) are: **because**, **since**.
    - Some of the key words that tell you what happened (effect) are: **therefore**, **as a result**.

Example Sentences

Here is an example of a cause and effect sentence:

- **Because** gold was discovered in the west (cause), many people moved to the west to get wealthy (effect).

Sometimes the order is reversed and the effect comes before the cause. Here is an example of an effect and cause sentence:

- Many people moved to the west (effect) **because** gold was discovered (cause).

Step 3 — Complete the reading map. Use the reading map to help you think about cause and effect.

Activity 16a — Read on Target for Grade 3

Map 16a

Cause and Effect
I read to find out the reason why something happened and what happened.

Mrs. Jones found a basket of three rabbits on her front step.

Cause: **Why** something happened (Reason/Action)

As a result →

Effect: **What** happened (result)

Activity 16a   *Read on Target* for Grade 3

Step 4

Read the following questions and write your answers.

1. Tell one thing that happened as a result of Mrs. Jones finding a basket of rabbits on her doorstep.

   _____
   _____
   _____

2. What is one thing Mrs. Jones did to let people know about the rabbits?

   _____
   _____
   _____

3. Why did Mrs. Jones decide to give away the rabbits?

   _____
   _____
   _____

4. What was the effect (result) of Mrs. Jones's yard sign?

   _____
   _____
   _____

Activity 16b

Identify Cause and Effect

I read to find out the reason why something happened and what happened.

Step 1

Read the article "Gold Rush Fever."

Gold Rush Fever

Before 1848, the West was mostly natural wilderness and frontier. But in 1848, gold was found at John Sutter's sawmill. News of the gold spread quickly in the California Territory. The news of the gold strike traveled to Missouri, where many people lived at the edge of the frontier. The excitement of the stories about miners getting rich swept through the country. President Polk reported to the nation that the gold strike was real. Soon, people from all parts of the country and even the world were headed to the California Territory to strike it rich.

In the 1840s, the trip to California was long and dangerous. Some people felt it was too dangerous for women and children. Most of the miners left their families behind to go and search for gold. At first, the miners lived in tents or shacks. As more miners came to search for the gold, more gold was found. The miners who found gold had money to spend. In time many of them built houses and sent for their families to join them. As more houses were built, the settlers came to live in the new towns. Now the miners had homes. California was growing. Soon, it would become a state.

Toward the end of 1856, the mines were running out of gold for miners to find. The new towns were overcrowded with newcomers. Because of the dream of finding more gold, many miners decided to stay in the new states where their dreams had brought them to live.

Activity 16b   *Read on Target* for Grade 3

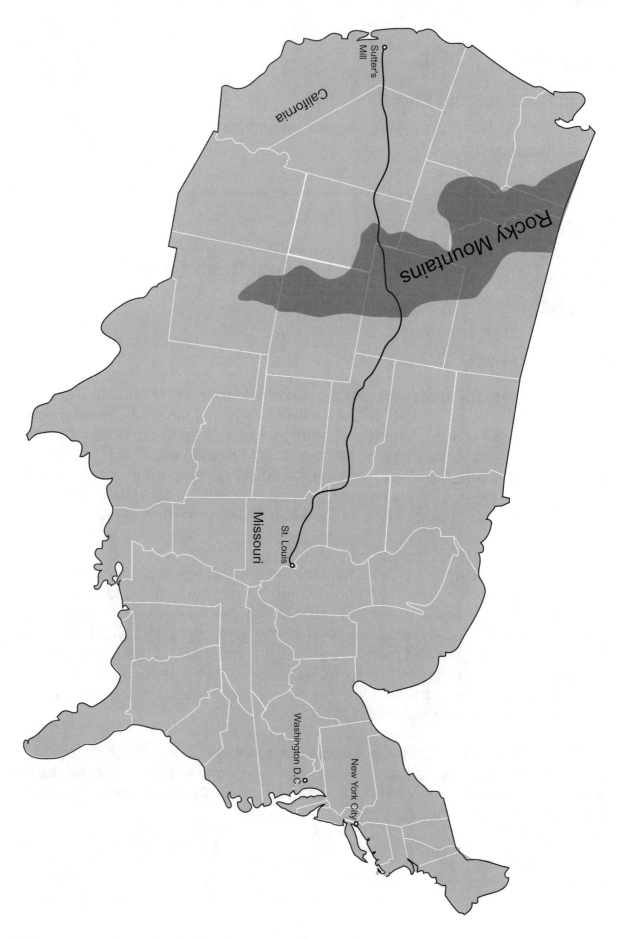

Step 2 — **Student Tips**

To identify cause and effect, you need to remember:

- The **cause** is the action or the reason that makes something happen (why something happened).
- The **effect** is the result (what happened).
- Look for key words that are clues to help you figure out if the statement is a cause or an effect.
  - Some of the key words that tell you why something happened (cause) are: **because**, **since**.
  - Some of the key words that tell you what happened (effect) are: **therefore**, **as a result**.

Example Sentences

Here is an example of a cause and effect sentence:

- **Because** gold was discovered in the west (cause), many people moved to the west to get wealthy (effect).

Sometimes the order is reversed and the effect comes before the cause. Here is an example of an effect and cause sentence:

- Many people moved to the west (effect) **because** gold was discovered (cause).

Step 3 — **Complete the reading map. Use the reading map to help you think about cause and effect.**

Activity 16b   Read on Target for Grade 3

Map 16b

Cause and Effect
I read to find out the reason why something happened and what happened.

[Effect: **What** happened (result)]

Gold was found in California in 1848.   As a result

Cause: **Why** something happened (Reason/Action)

[Effect: **What** happened (result)]

120   COPYING IS PROHIBITED   © 2005 Englefield & Associates, Inc.

Activity 16b   Read on Target for Grade 3

Step 4
Read the following questions and write your answers.

1. What caused most of the miners to leave their families behind and make the dangerous journey to the California Territory in 1848?

   _____

   _____

   _____

2. As the miners found more gold, how did they spend some of their money?

   _____

   _____

   _____

3. Why did the miners stay in California when the mines began to run out of gold?

   _____

   _____

   _____

4. How did miners finding gold help cause California to become a state?

   _____

   _____

   _____

Activity 16c

Identify Cause and Effect
I read to find out the reason why something happened and what happened.

Step 1 Read the story "My Horrible Day."

My Horrible Day

April 30th, my birthday, was a horrible, horrible day. It should have been a great day, but it wasn't. It could have been a great day, but it wasn't. There were two reasons I had a horrible day, and they were both my fault.

First, I stayed up the night before watching a television show that wasn't even very good. My bedtime is 9:00 because I have to wake up at 6:00 in the morning to be ready for school on time. I stayed up until 10:00, knowing that it would be really hard for me to wake up the next day and knowing that I am a really grumpy person if I don't get enough sleep.

Second, I didn't lay out my clothes for the next day. I always make sure I have a clean outfit the day before. I just didn't think about it when I was watching the show, and I was too tired after it was over.

These might not seem like big reasons to you, but they were the ruin of my birthday. I overslept because I was so tired that I didn't hear my alarm. I got up so late that my mother was yelling at me to get up and to hurry. Like I said, I am not a very nice person when I don't get enough sleep. I talked back to my mother and she grounded me. I screamed, "How can you ground me on my birthday?"

I didn't know that my mom hadn't done laundry yet, and all my clothes were either in the dirty clothes basket or on the floor of my bedroom. I had to get dressed out of the dirty clothes basket. My dad had already left for work, so he couldn't drive me to school after I missed my bus. My teacher and the other students had planned a surprise treat of cupcakes in homeroom to celebrate my birthday, but they ate them anyway when I didn't show up. Finally, my mom dropped my little brother off at his school (which started later than my school) and then dropped me off on her way to work. Since I didn't have a good excuse for being late, I had to stay in at lunch for being late to school. Then, one of my classmates pointed out the mustard stain on my shirt that I had gotten at lunch the other day. I felt grumpy, dirty, and embarrassed. Some birthday it turned out to be!

Step 2: Student Tips

To identify cause and effect, you need to remember:

- The **cause** is the action or the reason that makes something happen (why something happened).
- The **effect** is the result (what happened).
- Look for key words that are clues to help you figure out if the statement is a cause or an effect.
    - Some of the key words that tell you why something happened (cause) are: **because**, **since**.
    - Some of the key words that tell you what happened (effect) are: **therefore**, **as a result**.

Example Sentences

Here is an example of a cause and effect sentence:

- **Because** gold was discovered in the west (cause), many people moved to the west to get wealthy (effect).

Sometimes the order is reversed and the effect comes before the cause. Here is an example of an effect and cause sentence:

- Many people moved to the west (effect) **because** gold was discovered (cause).

Step 3: Complete the reading map. Use the reading map to help you think about cause and effect.

Activity 16c Read on Target for Grade 3

Map 16c

Cause and Effect
I read to find out the reason why something happened and what happened.

The narrator had a horrible day.

Effect: **What** happened (result)

Because

Cause: **Why** something happened (Reason/Action)

Cause: **Why** something happened (Reason/Action)

© 2005 Englefield & Associates, Inc.   COPYING IS PROHIBITED   125

Activity 16c Read on Target for Grade 3

Step 4

Read the following questions and write your answers.

1. What was the cause of the narrator's oversleeping?

   _____
   _____
   _____

2. What was the cause of the narrator being grounded by his mother?

   _____
   _____
   _____

3. What was one effect of the narrator's forgetting to set out his clothes the night before?

   _____
   _____
   _____

4. What was the overall effect of the narrator's oversleeping and forgetting to put out his clothes the night before?

   _____
   _____
   _____

Self-Scoring Chart

Rate how well you understand the critical-thinking steps by putting a star (★) for mastery, a plus sign (+) for making progress, and a minus sign (–) for needs help. You can rate what you know four different times.

Characters	1	2	3	4
Name a character				
Match sentences with the descriptions of the character				
Tell how the character impacts the story				
Tell how the story would be different if you changed a characteristic of the character				

Setting	1	2	3	4
Describe the setting: Tell where the story takes place				
Describe the setting: Tell when the story takes place				
Describe the setting: Tell what the setting looks like				
Tell how the setting affects the story				
Tell how the setting affects the events				
Change the setting: Change where the story takes place				
Change the setting: Change when the story takes place				
Change the setting: Change what the setting looks like				
Tell how the characters would be different if the setting changed				
Tell how the events would be different if the setting changed				

Plot	1	2	3	4
Describe the chain of events by writing the major events in correct order				
Change the plot by choosing an event to happen earlier or later				
Tell how the story is different when one of the events has been changed				
Take an event out of the story				
Tell what would be different if one of the events is left out of the story				
Write what would happen if the character's actions were different				

Self-Scoring Chart   Read on Target for Grade 3

Problem/Solution	1	2	3	4
Know the definition of a problem and a solution				
Write the problem of the story you read				
Write the events that lead up to the solution				
Write the solution of the story				
Write a different problem by making up your own problem				
Write how the events would be different				
Write how the solution would be different				

Point of View	1	2	3	4
Know the definition and key word pronouns of each point of view				
Identify the point of view in the story				
Write sentences from the story that helped you figure out the point of view				
Tell why the author writes from the point of view				
Tell how changing the point of view would affect the story				

Theme	1	2	3	4
Name some ideas from the text that tell what the story is about				
Write the lessons that the character learned				
Write a sentence telling what the message of the story is				

Infer	1	2	3	4
Read each sentence and paragraph to find clues about the story's meaning				
Write a clue in the clue box				
Write about an experience or knowledge you have of a similar thing				
Put the clue and your experience or knowledge together to make an inference about what is happening in the story				
Read more of the story to see if the inference is correct				

Self-Scoring Chart

Read on Target for Grade 3

	1	2	3	4
Predict				
Read each sentence and paragraph to find clues about the story's meaning				
Write a clue in the clue box				
Write about an experience or knowledge you have of a similar thing				
Put the clue and your experience or knowledge together to make a prediction about what will happen next				
Read more of the story to see if the prediction is correct				

	1	2	3	4
Compare and Contrast				
Write the names of the things to be compared and contrasted				
Describe the characteristics				
If the items have same characteristics, mark them with a plus sign (+)				
If the items have different characteristics, mark them with a minus sign (–)				

	1	2	3	4
Fact and Opinion				
Write sentences that tell you if the text is a fact or an opinion				
Write how the information can be proven by evidence or observation				
Write where you would look up information to check it				
Tell if the information is true for everyone				
Write key words that are clues to tell how someone thinks or feels				
Write how the information tells a personal belief or judgment				
Tell if the information is true for some people				

	1	2	3	4
Explain Purpose for Writing				
Know the definitions and purposes of fiction, poetry, and nonfiction				
Identify the type of writing				
Write the author's purpose				
Write a sentence or sentences from the story that show an example of the author's purpose				

Self-Scoring Chart — Read on Target for Grade 3

	1	2	3	4
Main Idea/Supporting Details				
Know the definitions of main idea and of supporting sentences				
Write the general topic of a paragraph				
Write one sentence that tells the main idea (overall message about the topic)				
Write detail sentences (to support, explain, or give information)				
	1	2	3	4
Respond to the Text				
Read the text and write your thoughts and feelings of a similar thing				
Write about your experiences or observations of a similar thing				
	1	2	3	4
Evaluate the Text				
Read the question given by the teacher				
List the strengths of the text				
List the weaknesses of the text				
After reading the strengths and weaknesses, answer the question				
	1	2	3	4
Summarize the Text				
Write the important information from each paragraph				
Rewrite each paragraph using your own words				
Use your own words to tell the overall idea of the whole selection				
	1	2	3	4
Identify Cause and Effect				
Know the definitions of cause and effect				
Read the key words for cause and effect				
Write why something happened (the cause)				
Write what happened (the effect)				

Notes

Notes

Notes

Notes

Notes

Subject-Specific Skill Development
Workbooks Increase Testing Skills

Write on Target
for grades 1/2, 3, 4, 5, and 6

Includes Graphic Organizers

Read on Target
for grades 1/2, 3, 4, 5, and 6

Includes Reading Maps

Math on Target for grades 3, 4, and 5

Includes Thinking Maps

For more information, call our toll-free number: 1.877.PASSING (727.7464)
or visit our website: www.showwhatyouknowpublishing.com